D1707532

Air Commandos

BY LINDA BOZZO

Amicus High Interest is an imprint of Amicus
P.O. Box 1329, Mankato, MN 56002
www.amicuspublishing.us

Library of Congress Cataloging-in-Publication Data
Bozzo, Linda.
Air commandos / by Linda Bozzo.
pages cm
Includes bibliographical references and index.
Summary: "An introduction to the life of Air Commandos in the Air Force (aka Air Force Special Operations Command (AFSOC)) describing some missions, how they train, and their role in the armed forces"– Provided by publisher.
Audience: Grades K-3.
ISBN 978-1-60753-489-1 (library binding) –
ISBN 978-1-60753-632-1 (ebook)
1. United States. Air Force–Juvenile literature. 2. Special operations (Military science)–Juvenile literature. 3. Special forces (Military science)–United States–Juvenile literature. I. Title.
UG633.B6964 2015
358.4–dc23

2013032337

Editor: Wendy Dieker
Series Designer: Kathleen Petelinsek
Book Designer: Steve Christensen
Photo Researcher: Kurtis Kinneman

Photo Credits:
Aero Graphics, Inc./CORBIS, cover; US Air Force Photo/Alamy, 5; KIM KYUNG-HOON/Reuters/Corbis, 6; FRANCIS R. MALASIG/epa/Corbis, 9; US Navy Photo/Alamy, 11; US Air Force Photo/Alamy, 12; U.S. Department of Defense/Science Faction/SuperStock, 15; US Air Force Photo/Alamy, 16/17; UK MOD Image/Alamy, 19; TBD 20; Biosphoto/SuperStock, 23; US Air Force Photo/Alamy, 24/25; U S Air Force photo/Staff Sgt Jonathan Snyder/SuperStock, 27; US Marines Photo/Alamy, 28/29

Printed in the United States at Corporate Graphics in North Mankato, Minnesota.
10 9 8 7 6 5 4 3 2 1

Table of Contents

A Call for Help

It is March 11, 2011. An earthquake strikes Japan. A **tsunami** follows. Homes are destroyed in the storm and flood. Thousands of people are hurt. Some are missing. Many more are dead. Air Commandos are called to help. Their mission: reopen runways at airports so planes can bring in supplies. The first teams fly over to look at the damage.

Soldiers in a chopper look at the area.

Air Commandos use equipment to clear the debris.

What kinds of supplies do people need?

The Commandos can't find a clear runway. They are all blocked with **debris**. But choppers don't need a runway to land. The Commandos quickly load gear on their choppers. Teams will work to clear a runway. The Commandos need to get to work fast. But a storm comes up. The choppers can't fly safely.

Storm victims need food, water, and blankets. They also need fuel for generators.

The Commandos need to use planes to get in. A runway needs to be cleared. Japanese soldiers work around the clock. Finally, a runway is clear. Planes are able to land. Equipment is brought in. Much needed supplies and workers can start to arrive. Air Commandos were able to bring hope back to the people of Japan.

Planes can bring supplies to help.

Learning the Ropes

Air Commandos are members of the Air Force Special Operations Command (AFSOC). Commandos start as **airmen**. They go through more training to join the AFSOC. They are the best of the best. Air Commandos bring extra power to the air force. The mission might be in the air. It could be **combat** on the ground. Air Commandos are strong. They are tough.

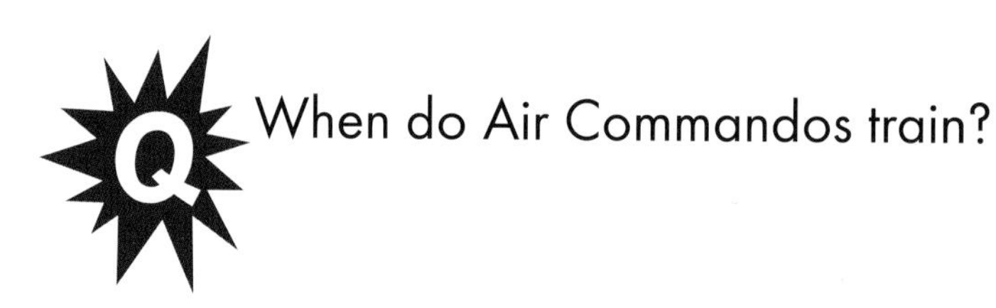

When do Air Commandos train?

Air Commandos climb out of choppers.

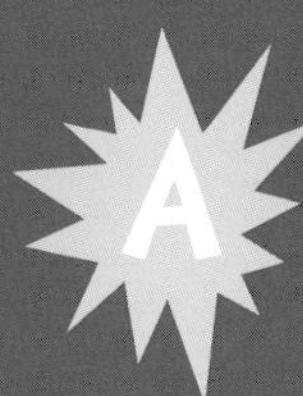

At night. Most Air Commando missions happen at night, so they practice in the dark.

This soldier trains to dive into water.

As airmen, Commandos train in the air. Commandos learn to fly planes and choppers. They even learn how to jump from aircraft. But they also train on land and in the water. They ride motorcycles and snowmobiles. They swim and ski. Some learn to use **radar** and **satellites** to study the weather and the land.

The skills that Commandos learn can make the difference between life and death. Many Commandos have **medical** training. They learn how to use all kinds of **weapons**. Some will learn to blow up bombs. All Commandos learn what to do if the enemy catches them. They learn how to survive. They learn how to escape.

Soldiers learn to hide from the enemy.

Would you do anything to save a person's life? Even go behind enemy lines? What about jump from a plane? Maybe try a daring rock-climbing rescue? There are many jobs Air Commandos can train for. What is their most important job? To always be ready for the next mission.

Commandos jump from planes to rescue people.

The Home Front

On the home front, Air Commandos train. They plan missions. They fix vehicles. They pack supplies. Air Commandos prepare for all kinds of jobs. They are ready at all times for new missions. The home front is also a place for airmen to recover after a mission overseas.

Fixing aircraft can be a job for a Commando.

Commandos bring injured people to safety.

At home, Air Commandos work in the community. They do volunteer work. They visit schools. Search and rescue teams help after storms. They also help fix things that were damaged.

Air Commandos are there when trouble strikes. They help stop **illegal drugs** from coming into our country. These soldiers help keep us safe.

Stationed Overseas

Air Commandos carry out missions overseas. For weeks, they work in the middle of nowhere in secret. Quick. Quiet. They work in all kinds of weather. Most of their missions are at night. This is so they cannot be seen. Commandos are close to war overseas. The enemy is never far away. Danger is near.

How can Air Commandos see in the dark on their night missions?

Soldiers must move quickly and quietly. They can't be seen.

A

Air Commandos use special goggles called panoramic night vision goggles (PNVGs). These help them see in the dark.

Air Commandos risk their lives to save others. Small teams get in. They do their jobs and get out. They provide ground and air refueling. Teams rescue troops trapped or hurt behind enemy lines. They destroy the enemies. Troops cannot always be reached by roads. Commandos fly choppers to bring in **ammunition** and supplies.

Commandos help people who are hurt.

Air Commandos help other special forces too. They bring extra fighting power. Their job is to do what other forces cannot. Small, skilled teams work quickly. They are trained to use special equipment that pinpoints enemies. They can rescue people. Whatever it takes, they get the job done.

Do other branches of the military have special forces?

Air Commandos work in small teams on missions.

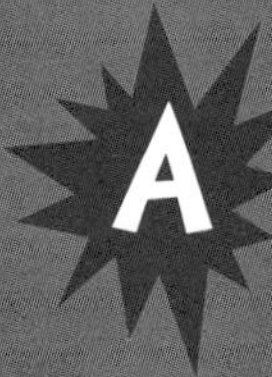

Yes! Each branch has units of highly trained soldiers. Some others are the Green Berets, Delta Force, and the Navy SEALs.

Serving Our Country

Air Commandos serve our country by protecting our freedom. Training troops in other countries helps to fight **terrorism**. Commandos keep their eyes on the skies. They watch for the enemy. They are ready for any mission. This is how they help save lives. Their motto is: "Any Time, Any Place."

Supplies are dropped for soldiers in combat.

Glossary

airmen Trained air force recruits.

ammunition Bullets fired from weapons.

combat Missions that include fighting in a war.

debris Scattered pieces of something that has been destroyed.

illegal drugs Drugs that are against the law.

medical To do with doctors or medicine.

radar An instrument that finds aircraft using radio waves.

satellite Equipment in space that is used to gather information.

terrorism An attack that causes fear.

tsunami A huge crushing wave from an underwater earthquake or volcano.

weapon An object used in a fight to attack or defend.

Read More

Gordon, Nick. *Air Force Air Commandos.* Minneapolis: Bellwether Media, 2013.

Lusted, Marcia Amidon. *Air Commandos: Elite Operations.* Minneapolis: Lerner Publications, 2014.

Williams, Brian. *Special Forces.* Chicago: Heinemann Library, 2012.

Websites

Air Force Special Operations Command: Air Commandos
www.afsoc.af.mil/aircommandos.asp

Brain Pop: Armed Forces
www.brainpop.com/socialstudies/usgovernmentandlaw/armedforces/preview.weml

U.S. Air Force
www.af.mil/

Every effort has been made to ensure that these websites are appropriate for children. However, because of the nature of the Internet, it is impossible to guarantee that these sites will remain active indefinitely or that their contents will not be altered.

Index

About the Author

Linda Bozzo is the author of more than 45 books for the school and library market. She would like to thank all of the men and women in the military for their outstanding service to our country. Visit her website at www.lindabozzo.com.